WAVE-PARTICLE

DUALITY

How & Why a Photon form a Wave?

BY

NARENDRA SWARUP AGARWAL

Copyright © 2021 Narendra Swarup Agarwal

All rights reserved.

ISBN-13: 9798705850358

DEDICATION

Dedicated to my parents who taught me to observe and analyse nature.

PREFACE

The Light or the Electromagnetic Waves consist of particles known as photons. Even though a photon is a particle, it always moves as a wave and display the characteristics of both the particle and the wave, known as Wave-Particle Duality.

The scientists could not explain the dual nature of the photons during the last few centuries and accepted Wave-Particle Duality as the fundamental nature of the photons.

Only the New Quantum Theory developed in the year 2012 in India solved the mystery of Wave-Particle Duality.

This book explains Wave-Particle Duality in detail. How and Why a Photon forms a Wave and displays the behaviour of both the particle as well as wave?

TABLE OF CONTENTS

1.0	Introduction	1
2.0	New Quantum Theory	5
3.0	The Structure of a Photon	9
3.0	Mass of a Photon	16
4.0	Wave-Particle Duality	23
4.1	The Mechanical Forces	25
4.2	Wave-Particle Duality Explained	41
4.3	To Find Exact Location of a Photon	64

WAVE-PARTICLE DUALITY

CHAPTER 1.0

INTRODUCTION

The scientists worked on light for more than 400 years and developed theories of light. In 1637, Descartes developed the 'Corpuscular Theory of Light' which was elaborated by Isaac Newton in 1672. According to this theory, the light is made up of small discrete particles called 'Corpuscles' (small particles) which travel in straight line with a finite velocity.

Christiaan Huygens proposed a mathematical 'Wave Theory of Light' in 1678 and published in his Treatise of Light in 1690.

Huygens proposed that light was emitted in all directions as a series of 'Waves' in a medium called the Luminiferous Ether vibrating up and down.

The double slit experiment by Thomas Young in 1801 proved the Wave Theory of Light. The actual distribution of brightness can be explained by the alternately additive and subtractive interference of waves. But the light is absorbed as the particles on the screen.

Since the 17th century, the confusion prevailed 'Whether the light is a particle or a wave?'

The double slit experiment proved the dual nature of the light, the Wave-Particle Duality. The light exhibits the behaviour of both particles as well as wave.

Max Planck, Albert Einstein and the combined efforts of several eminent scientists in the 20th century developed the Quantum Theory. However, the Dual nature of the Photons could not be explained and remained a mystery.

WAVE-PARTICLE DUALITY

In absence of any solution, the scientists in the 20th century accepted Wave-Particle Duality as a fundamental property of the photons.

But Einstein did not accept the Wave-Particle Duality and wrote:

"This double nature of radiation (and of material corpuscles) is a major property of reality, which has been interpreted in quantum mechanics in an ingenious and amazingly successful fashion. This interpretation, which is looked upon essentially as final by almost all contemporary physicists appears to me as only a temporary way out."

The Wave-Particle nature of Light or the Electromagnetic Waves or the dual nature of Light remained a puzzle for the scientists and could not be explained till recently.

ONLY THE NEW QUANTUM THEORY, DEVELOPED BY THE AUTHOR IN THE YEAR 2012 IN INDIA EXPLAINS WAVE-PARTICLE DUALITY.

Narendra Swarup Agarwal

CHAPTER 2.0

NEW QUANTUM THEORY

The New Quantum Theory is based on the universal facts.

Everything around us is made up from very tiny atoms. An atom has a nucleus of the mass which is more than 99.9% mass of the total mass of the atom, but the nucleus occupies a negligibly small volume, less than 10^{-10} of the total volume of the atom. Moreover, the nucleus of an atom is not in the centre of the atom but located off-centre. The nucleus of the atom has a charge.

Similarly, on an exceptionally large scale, the solar system has the Sun as a nucleus. The mass of the Sun is more than 99.86% mass of the total mass of the solar system. The Sun occupies only negligibly small volume less than 10^{-10} of the total volume of the solar system.

All the planets in the solar system orbit around the Sun in the elliptical paths. The Sun is not in the centre of any of the elliptical path but at one of the epicenters. Therefore, the Sun is also not in center of the solar system but occupies an off-centre position in the solar system. The Sun is the most significant source of Electromagnetic radiations and has the charge. The tiny particles such as atom and the exceptionally large solar system have similarities.

The **NEW QUANTUM THEORY** is based on these universal facts and states:

"A photon/quantum particle has a nucleus of the mass and the charge located off-centre. The nucleus has nearly all the mass and charge of the photon."

WAVE-PARTICLE DUALITY

A small nucleus of the mass and the charge, located off-centre in the spinning photon/quantum particle, bestows unique characteristics to display strange phenomena.

This constitutional essence of a photon/quantum particle explains not only the Wave-Particle Duality but all the Quantum Phenomena and completes the Quantum Theory.

Narendra Swarup Agarwal

CHAPTER 3.0

THE STRUCTURE OF A PHOTON

It is already known that the photons are spinning particles and move with velocity of light as a wave and display mysterious quantum phenomena. The structure of a photon based on the New Quantum Theory is explained below:

Figure 1 shows a Photon with a nucleus of mass and charge which is not in centre of the Photon.

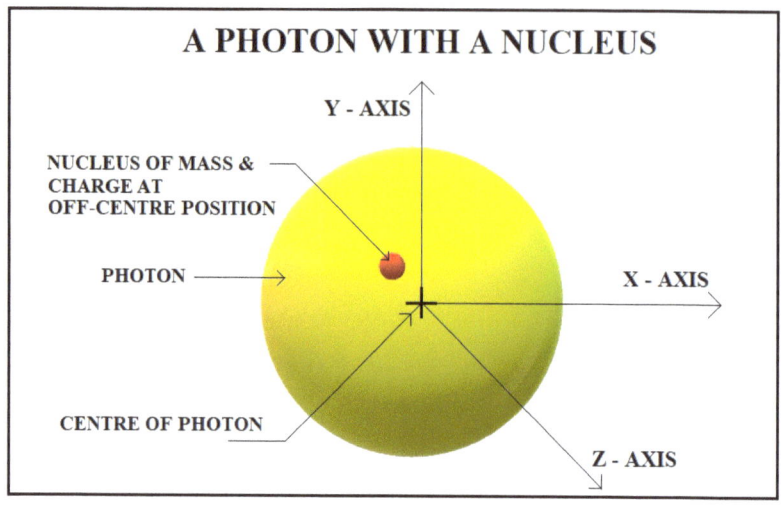

Figure 1: The big yellow sphere shows a photon, and the small red sphere shows a nucleus of mass and charge in the photon. The nucleus is not in the centre of photon but located off-centre in the photon.

WAVE-PARTICLE DUALITY

Figure 2 shows a 3D view of a spinning photon.

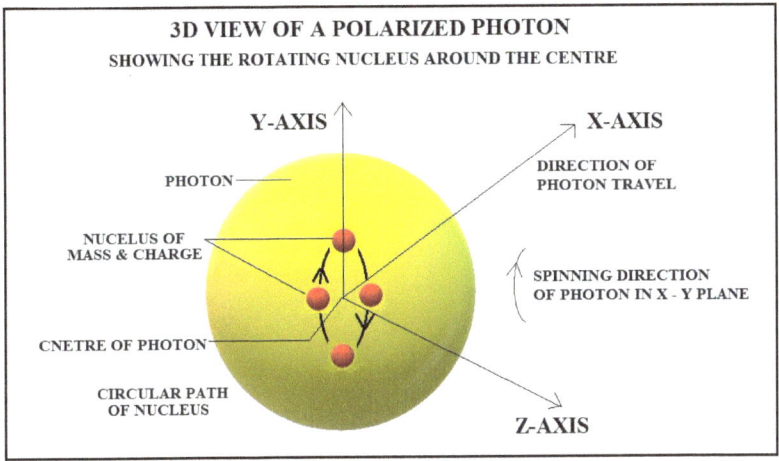

Figure 2: The big yellow sphere shows a spinning photon. The red sphere is the nucleus of mass and charge. The red nucleus rotates around the centre of the photon. The Figure 2 shows a few different positions of the red nucleus in the circular path rotating around the centre of the photon.

The spinning photon has only one nucleus of mass and charge as the red sphere. As the photon spins, the nucleus of mass and charge rotates around the centre of the photon.

Figure 3 shows a cross section of the photon through its centre and the nucleus of mass and charge.

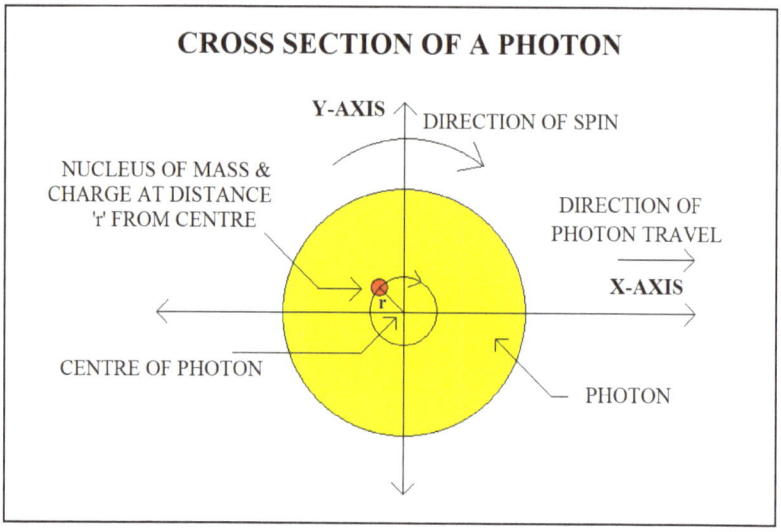

Figure 3: The yellow circle shows the cross section of a photon. The small red circle is the cross section of nucleus of mass and charge. The distance between the nucleus and centre of photon is 'r'.

As the photon spins, the nucleus rotates around the centre of the photon in a circle of radius 'r'. The photon in the figure spins in a clockwise direction and moves in the X-direction.

Figure 4 shows an enlarged cut section of only the central portion of the photon through its centre and the nucleus of the photon.

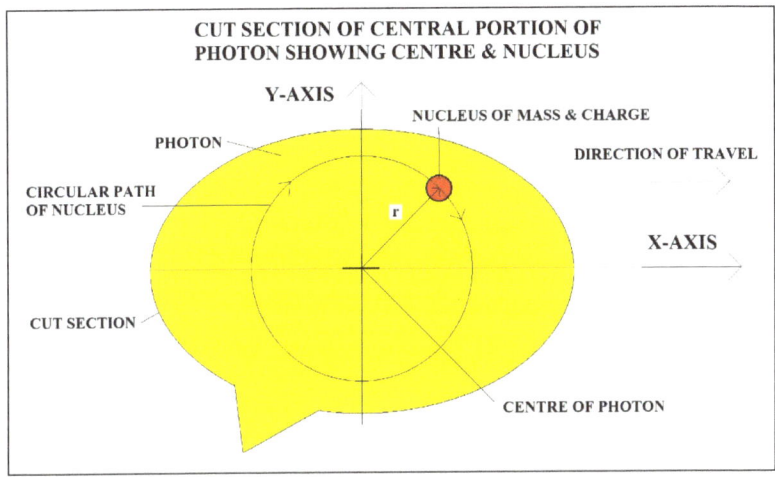

Figure 4: An enlarged Cut Section of the central portion of a Photon showing the nucleus of mass and charge and its circular path around the centre of the photon.

In a photon of frequency 'f' cycles per second, the nucleus also completes 'f' revolutions around the centre of the photon in one second.

The 'New Quantum Theory' describes the structure of a photon as under:

- A photon has a nucleus of mass and charge. The nucleus has practically all the mass and the charge of the photon.
- A photon has a nucleus in its off-centre position at a distance 'r' from the centre of the photon.
- As the photon spins the nucleus of mass and charge rotates in a circle around the centre of the photon in a circle of radius 'r'.
- The nucleus of mass and charge of the photon revolves 'f' number of times around the centre of the photon in one second, where 'f' is the frequency of the photon.
- The above structure differentiates between linearly and circularly polarized photon as under:
- A photon spins in only one plane, say X-Y plane. If the photon travels in any direction which confines in X-Y plane, the photon is linearly polarized photon.

WAVE-PARTICLE DUALITY

- A photon spinning in X-Y plane and traveling in X-Axis is linearly polarized photon and generates a wave which confines in X-Y plane only.
- A photon spinning in X-Y plane and traveling in any direction which does not match X-Y plane, the photon is circularly polarized photon.
- Such a photon generates a helix wave in all X-Y-Z planes.
- A photon spins with constant frequency 'f', the nucleus of mass and charge rotates at a constant linear speed around the centre of the photon in X-Y plane.
- However, the velocities of the nucleus in both the X-Axis and Y-Axis continuously accelerate or decelerate with the rotation of the nucleus.
- The continuous acceleration/deceleration of the mass generates forces of varying magnitude. Similarly, the continuous acceleration/deceleration of the charge generates Electromagnetic forces.

CHAPTER 3.1

MASS OF A PHOTON

The Planck-Einstein equation $E = hf$ specifies the energy 'E' in all 'PHYSICAL FORMS' of energy possessed by a photon.

Note: This energy 'E' does not include the energy 'e' of the mass-energy equivalence reaction $e = m\,c^2$ which is applicable only when the mass of the photon is converted to energy as already explained.

WAVE-PARTICLE DUALITY

The Mass of photon 'm' can be calculated from the energy of the photon as under:

Energy of Photon (E) = Linear Kinetic Energy + Rotational Kinetic Energy

Linear Kinetic Energy of a photon : $\frac{1}{2}mc^2$

Rotational Kinetic Energy of a photon : $\frac{1}{2}I\omega^2$

Energy of a photon :

$$E = \frac{1}{2}mc^2 + \frac{1}{2}I\omega^2$$

Or $$E = hf = \frac{1}{2}mc^2 + \frac{1}{2}mr^2(2\pi f)^2$$

Mass of a Photon $$m = \frac{2hf}{c^2 + 4\pi^2 r^2 f^2}$$

In the above equation, the term $4\pi^2 r^2 f^2$ in the denominator is negligibly small in comparison to c^2.

Therefore, the equation for the Mass of Photon is simplified to:

Mass of Photon $$m = \frac{2hf}{c^2}$$

Or $$m = af$$

Where 'a' is constant with a name 'Mass - Frequency Constant'. The value of this constant is given below:

$$a = \frac{2h}{c^2} = 1.474499440284 \times 10^{-50}$$

Where:

m: Mass of the photon

c: Linear velocity of the photon in X-direction

I: Moment of Inertia of the photon

ω: Angular velocity of the photon

f: frequency of the photon

WAVE-PARTICLE DUALITY

r: Distance of centre of mass of the photon from the centre of the photon

h: Planck's constant

a: A new constant known as 'Mass - Frequency Constant'

Notes:

1. The value of '*r*' being of the order of 10^{-17} meter and the term $4\pi^2 r^2 f^2$ is negligibly small in comparison to c^2 which is of the order of 10^{16} meters.

2. The potential energy is not considered here considering no or negligible change in the elevation.

THE MASS OF A PHOTON IS CONSTANT. IT DOES NOT CHANGE UNLESS AND UNTIL THE PHOTON INTERACTS WITH SOME OTHER OBJECT.

The terms Rest Mass or Relativistic Mass etc. of a photon are only the conditional interpretations.

Mass of a Photon of frequency 'f':

$$m = af$$

or $\quad m = 1.474499440284 \times 10^{-50} f$

The 'Mass of a photon' of a known frequency can be calculated simply by multiplying the frequency of the photon with 'a' (Mass - Frequency constant).

$$a = \mathbf{1.474499440284 \times 10^{-50} \text{ kg sec.}}$$

The mass of a Red photon is 6.98×10^{-36} and Dark Blue photon is 1.000×10^{-35} Kg.

Mass of a few Photons of different wavelengths/frequencies are calculated in **Table 1**.

WAVE-PARTICLE DUALITY

Table No. 1

Mass of Photon for Different Wavelength & Frequency

Wavelength (nano meter)	Energy (Joule)	Type of Radiation	Frequency (f) (cycles/second)	True Mass of Photon (af) (Kg)
0.001	$1.9864458 \times 10^{-13}$	Gamma Rays	$2.99792458 \times 10^{20}$	$4.420438115 \times 10^{-30}$
1.0	$1.9864458 \times 10^{-16}$	X Rays	$2.99792458 \times 10^{17}$	$4.420438115 \times 10^{-33}$
100	$1.9864458 \times 10^{-18}$	Ultra-Violet	$2.99792458 \times 10^{15}$	$4.420438115 \times 10^{-35}$
442	$4.4942213 \times 10^{-19}$	Dark Blue	$6.78263479 \times 10^{14}$	$1.000099121 \times 10^{-35}$
531	$3.7409525 \times 10^{-19}$	Green	$5.64580900 \times 10^{14}$	$8.324742213 \times 10^{-36}$
633	$6.6260700 \times 10^{-19}$	Red	$4.73605778 \times 10^{14}$	$6.983314558 \times 10^{-36}$
1000	$1.9864458 \times 10^{-19}$	Infra-Red	$2.99792458 \times 10^{14}$	$4.420438115 \times 10^{-36}$
1,000,000	$1.9864458 \times 10^{-22}$	Ultra High Frequency	$2.99792458 \times 10^{11}$	$4.420438115 \times 10^{-39}$
1,000,000,000	$1.9864458 \times 10^{-25}$	Very High Frequency	2.99792458×10^{8}	$4.420438115 \times 10^{-42}$

CHAPTER 4.0

WAVE - PARTICLE DUALITY

A photon is a particle but has the unique characteristics to move as a wave. A particle moves up and down from its line of travel and forms a wave on its own without any external force.

The velocity and direction of the photon change continuously in one wave cycle of 360^0 spin of the photon and repeat in every wave cycle.

A force is necessary to change the velocity and direction of a particle.

This confirms that the mysterious particles of photons or quantum particles have unique capabilities to produce self-generated forces.

According to the New Quantum Theory, a photon or a quantum particle has unique structure, a NUCLEUS OF MASS AND CHARGE, located in off-centre position.

The nucleus of mass and charge in off-centre position, in the spinning photon, experiences continuous acceleration/deceleration. The mass on acceleration/deceleration generates forces and these continuously varying forces move the photon as a wave.

WAVE-PARTICLE DUALITY IS ONE OF THE DECISIVE PROVES OF THE PRESENCE OF MASS IN An OFF-CENTRE POSITION IN THE PHOTON.

CHAPTER 4.1

THE MECHANICAL FORCES

DEVELOPED BY A PHOTON

(INTERNAL FORCES FROM WITHIN THE PHOTON)

A photon develops continuously varying Mechanical Internal Forces in different directions to move the photon either as a helix, a 3D wave or a 2D wave in one plane.

As a photon spins, its nucleus of the mass and the charge rotates around the centre of the photon and rotates 360^0 angle around the centre of the photon in one spin of the photon.

PHASE ANGLE:

The angle of 'Nucleus of mass and charge' from the Axis/Direction of travel of the photon is known as 'Phase Angle of the photon or the nucleus'. The term phase angle may be well known to most of the readers, for the others it is explained below:

A new-wave cycle of the photon starts from X – Axis (0, 0) coordinates when the nucleus of mass and charge (red sphere) is at 0^0 phase angle from the X – Axis as shown in **Figure 5**.

WAVE-PARTICLE DUALITY

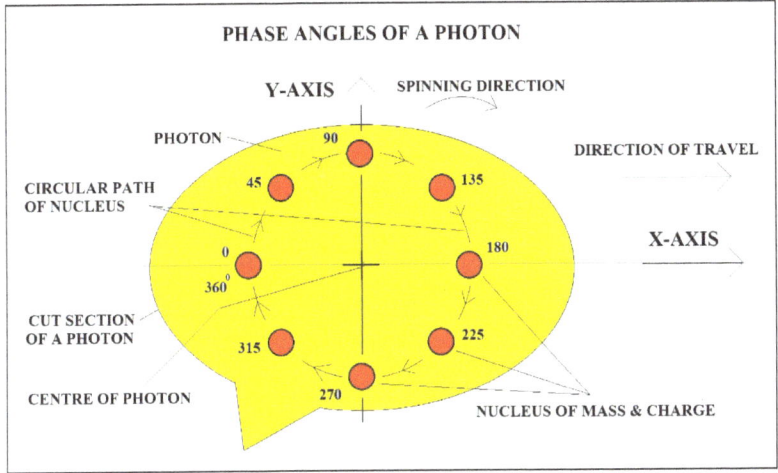

Figure 5: Cut section of the central portion of a linearly polarised photon as yellow circle is shown in the Figure. The photon travels in X-Axis. The red circles show the different phase angles of the nucleus (or the photon). The phase angles of the nucleus from 0^0 to 360^0 are shown with 45^0 increments.

When the nucleus is on the X-Axis on the left-hand side, the phase angle is 0^0. As the photon spins in the clockwise direction, the nucleus rotates, and the phase angle increases continuously and reaches to the maximum value 360^0.

MECHANICAL FORCES DEVELOPED FROM WITHIN A PHOTON:

A linearly polarised photon moves in X-Axis, forms a wave in X–Y plane, a 2D plane and develops the mechanical forces as under.

A photon spins with frequency 'f', moves in X-Axis and forms wave in the X-Y plane.

A photon spins in a plane consisting the nucleus of mass and charge of the photon and the centre of the photon.

The nucleus of mass and charge rotates in a circle around the centre of the photon completing 'f' revolutions per second.

The spinning photon travels in X – Axis and continuously moves up and down in Y – Axis to form a Wave in X – Y plane by its self-generated mechanical forces.

An enlarged view of the central part of the spinning photon with the nucleus of mass and charge is shown in **Figure 6**.

As the photon spins, the phase angle of the nucleus increases from θ to $(\theta + d\theta)$ by $d\theta$ angle in time dt.

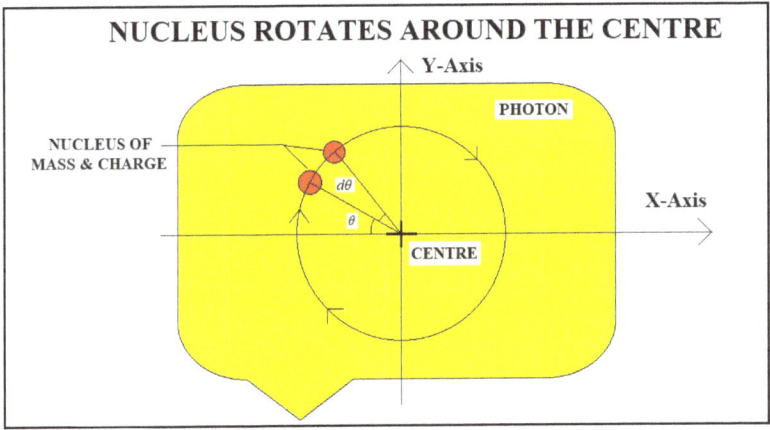

Figure 6: The nucleus rotates in the circular path and in time dt the phase angle increases from θ to $(\theta + d\theta)$ degree.

The frequency of a photon 'f' is constant and the photon spins with constant angular velocity.

The nucleus rotates inside the photon with the constant angular velocity and the constant linear velocity around the centre of the photon in a circular path.

As the photon spins in a X–Y plane consisting the centre of the photon and the nucleus of the photon, the velocities of the nucleus of mass and charge vary continuously in both X and Y directions.

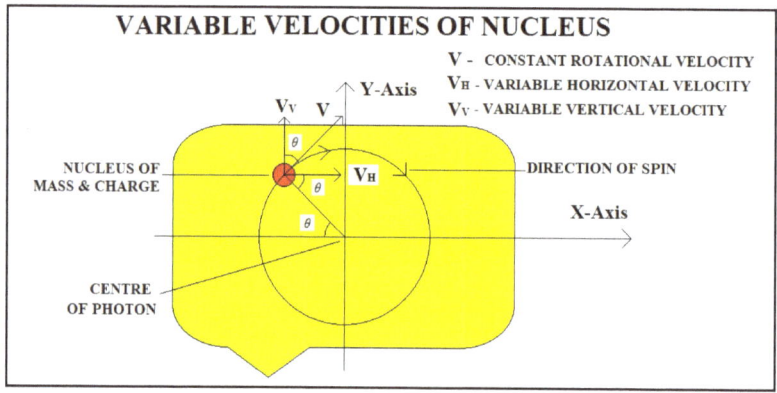

Figure 7: As the photon spins clockwise, the nucleus rotates around the center of the photon with constant linear velocity 'V'. The velocities of the nucleus in both horizontal and vertical directions vary continuously with the phase angle.

WAVE-PARTICLE DUALITY

The nucleus of a photon develops the following velocities and forces:

Circumference of the circle of nucleus: $2\pi r$ meter

Frequency of the photon: f cycles/sec

Constant Linear Velocity of the rotating Nucleus:

$$2\pi r f \text{ meter/sec}$$

For every new-wave cycle or one revolution of the photon, the phase angle of the nucleus increases from 0^0 to 360^0. The instantaneous velocities of the nucleus at θ^0 phase angle of the nucleus are calculated below:

Horizontal velocity in X-direction: $2\pi r f \sin\theta$

Vertical velocity in Y-direction: $2\pi r f \cos\theta$

With the spin of the photon, in time dt, the phase angle of the nucleus increases by $d\theta$ increasing the phase angle from θ to $\theta + d\theta$.

The instantaneous velocities, accelerations and forces developed at $\theta + d\theta$ phase angle of the nucleus are calculated as under:

Horizontal velocity in X - direction: $2\pi r f \sin(\theta + d\theta)$

Vertical velocity in Y - direction: $2\pi r f \cos(\theta + d\theta)$

Change in the Horizontal velocity:

$$2\pi r f \sin(\theta + d\theta) - 2\pi r f \sin\theta$$

Change in the Vertical velocity:

$$2\pi r f \cos(\theta + d\theta) - 2\pi r f \cos\theta$$

Acceleration a_x in X - Direction:

$$[2\pi r f \sin(\theta + d\theta) - 2\pi r f \sin\theta] / dt \ \text{m/sec}^2$$

Acceleration a_y in Y - Direction:

$$[2\pi r f \cos(\theta + d\theta) - 2\pi r f \cos\theta] / dt \ \text{m/sec}^2$$

WAVE-PARTICLE DUALITY

The above accelerations create the 'Forces' within the photon in both the directions 'X' and 'Y' due to the mass '*m*' in the nucleus.

Force F_x in X - direction:

$$m \, [2 \pi r f \sin(\theta + d\theta) - 2 \pi r f \sin \theta] / dt$$

or $2 \pi r f m \, [\sin(\theta + d\theta) - \sin \theta] / dt$ kg meter/sec^2

Force F_y in Y-direction:

$$m \, [2 \pi r f \cos(\theta + d\theta) - 2 \pi r f \cos \theta] / dt$$

or $2 \pi r f m \, [\cos(\theta + d\theta) - \cos \theta] / dt$ kg meter/sec^2

The above equations calculate the velocities, accelerations and the forces of a polarized photon in X and Y – directions at any given time or the phase angle of the nucleus of the photon in a Wave Cycle of the photon.

As the photon spins with a constants frequency and completes 360^0 spin in one revolution, both the phase angle and time have constant relationship as under:

Time for one revolution of photon: $1/f$ second

(The photon completes one spin of 360^0 phase angle)

Time for 1^0 spin of photon or: $1/360f$ second

(1^0 rotation of Nucleus around the centre of photon)

The Velocities at 0^0 phase angle due to the nucleus:

(At the starting point of a new wave cycle the phase angle θ is zero, therefore, Sin θ is zero and Cos θ is 1).

Horizontal velocity in X – direction: 0

Vertical velocity in Y - direction: $2\pi r f$ meter/sec

The above velocities are due to the presence of a mass in the nucleus of the photon.

WAVE-PARTICLE DUALITY

The Velocities at 90^0 phase angle of the photon:

(At phase angle θ is 90^0, Sin θ is 1 and Cos θ is zero).

Horizontal velocity in X - direction: $2\pi r f$ meter/sec

Vertical velocity in Y - direction: 0

The Velocities at 180^0 phase angle of the photon:

(At phase angle 180^0, Sin θ is zero and Cos θ is minus 1.

Horizontal velocity in X - direction: 0

Vertical velocity in Y - direction: (-) $2\pi r f$ meter/sec

The Velocities at 270^0 phase angle of the photon:

(At phase angle 270^0, is zero Sin θ is minus 1 and Cos θ is zero).

Horizontal velocity in X - direction: (-) $2\pi r f$ meter/sec

Vertical velocity in Y - direction: 0

The above velocities are due to the rotation of the nucleus in a circular path around the centre of the photon with the spin of the photon.

A photon also has its inherent velocity (velocity of light 'c') in X – direction, the net horizontal velocity at phase angle θ is given below:

Total Horizontal velocity in X - direction:

$$c + 2\pi r f \sin\theta \text{ meter/sec}$$

IMPORTANT NOTES:

- **The value of Sin θ is zero or has positive value during 0 to 180^0 phase angle or upper half of a wave cycle. Therefore, the horizontal velocity of a photon is MORE than the average velocity of light 'c' in the upper half of the wave cycle.**

WAVE-PARTICLE DUALITY

- The value of Sin θ is zero or has negative value during 180 to 360^0 phase angle or lower half of the wave cycle. Therefore, the horizontal velocity of a photon is LESS than the average velocity of light 'c' in the upper half of the wave cycle.
- In one full wave cycle of the photon, the average velocity of the photon is 'c' (velocity of light). The increase in velocity in 0^0 to 180^0 is nullified by the decrease in velocity in 180^0 to 360^0.

FORCES DEVELOPED BY A CIRCULARLY POLARIZED PHOTON

A circularly polarized photon forms a wave in all the 3 planes as a helix. The velocities, accelerations and forces in all the 3 directions X, Y & Z directions can be calculated using the similar equations given above.

The nucleus of mass and charge of the photon moves in all the 3 axes.

If θ_1 is phase angle in X-Axis and increases by $d\theta_1$ in time dt, the nucleus of the photon develops force in X-Axis as under:

Acceleration in X-Axis:

$$2\pi r f [\sin(\theta_1 + d\theta_1) - \sin\theta_1] / dt \text{ m/sec}^2$$

Force in X-Axis:

$$2\pi r f m [\sin(\theta_1 + d\theta_1) - \sin\theta_1] / dt \text{ kg meter/sec}^2$$

If θ_2 is phase angle in Y-Axis and it increases by $d\theta_2$ in time dt, the nucleus of the photon develops force in Y-Axis as under:

Acceleration in Y-Axis:

$$2\pi r f \, [\mathrm{Sin}\,(\theta_2 + d\theta_2) - \mathrm{Sin}\,\theta_2] / dt. \text{ m/sec}^2$$

Force in Y-Axis:

$$2\pi r f m \, [\mathrm{Sin}\,(\theta_2 + d\theta_2) - \mathrm{Sin}\,\theta_2] / dt \text{ kg meter/sec}^2$$

If θ_3 is phase angle in Z-Axis and it increases by $d\theta_3$ in time dt, the nucleus of the photon develops force in Z-Axis as under:

Acceleration in Z-Axis:

$$2\pi r f \, [\mathrm{Sin}\,(\theta_3 + d\theta_3) - \mathrm{Sin}\,\theta_3] / dt. \text{ m/sec}^2$$

Force in Z-Axis:

$$2\pi r f m \, [\mathrm{Sin}\,(\theta_3 + d\theta_3) - \mathrm{Sin}\,\theta_3] / dt \text{ kg meter/sec}^2$$

The above equations calculate the velocities of a photon in all the 3 directions X, Y & Z at any phase angle in the wave cycle. The phase angle is related to the time as the frequency of the photon is known.

The size and shape of a helix, a 3D wave of a circularly polarized photon can be calculated using the above information.

CHAPTER 4.2

WAVE - PARTICLE DUALITY EXPLAINED

(HOW A PHOTON FORMS A WAVE)?

Consider a linearly polarized photon moving in X–Axis with an average velocity 'c' and forming a 2D wave in X-Y plane.

The photon spins and its nucleus of mass and charge rotates around the centre of the photon in X-Y plane. The mass rotating around the centre of the photon develops varying velocities, accelerations and forces in both the X as well as Y – directions. The varying forces with the spin of the photon move the photon as a wave.

The following explains the Wave-Particle Duality phenomena:

- As the photon spins, the nucleus of mass and charge rotates around the centre of the photon.
- The linear speed of the rotating nucleus is constant. But its velocity vector changes continuously with the spin of the photon.
- The velocity in X – direction is the function of $Sin\theta$ and the velocity in Y – direction is the function of $Cos\theta$.

WAVE-PARTICLE DUALITY

- As the phase angle of the photon changes, the velocities vary in both the X and Y directions.
- The changes in the velocities create accelerations in both the X and Y directions.
- The accelerating mass in the nucleus of the photon develops the forces in both the X and Y directions.
- The velocities, accelerations and forces in both the X and Y directions vary continuously in one wave cycle of the photon.

As the photon spins, the phase angle of the nucleus of mass and charge increases from $0°$ to $360°$ in one wave cycle.

The instantaneous velocities of the nucleus at $\theta°$ phase angle of the nucleus are already calculated in the previous chapter and given below:

Horizontal velocity in X-direction: $c + 2\pi r f \sin\theta$

Vertical velocity in Y-direction: $2\pi r f \cos\theta$

The nucleus of the spinning photon develops the accelerations and forces during its rotation from θ to $(\theta + d\theta)$ phase angle of the nucleus in time dt second:

Acceleration a_x in X - Direction:

$$[2\pi r f \sin(\theta + d\theta) - 2\pi r f \sin\theta] / dt \text{ m/sec}^2$$

Acceleration a_y in Y - Direction:

$$[2\pi r f \cos(\theta + d\theta) - 2\pi r f \cos\theta] / dt \text{ m/sec}^2$$

Force F_x in X - direction:

$$2\pi r f m [\sin(\theta + d\theta) - \sin\theta] / dt \text{ kg meter/sec}^2$$

Force F_y in Y-direction:

$$2\pi r f m [\cos(\theta + d\theta) - \cos\theta] / dt \text{ kg meter/sec}^2$$

The values of $\sin\theta$ and $\cos\theta$ change between +1 and (−) 1. The forces created by the mass in the nucleus of the photon change directions in both X, (-) X, Y and (-) Y directions.

Therefore, a photon moves up and down in a wave cycle.

WAVE-PARTICLE DUALITY

THEREFORE, IN A WAVE CYCLE THE PHASE ANGLE OF THE NUCLEUS/PHOTON IS VERY IMPORTANT FOR THE CHARACTERISTICS OF THE PHOTON.

THE PHASE ANGLE IS VERY IMPORTANT NOT ONLY FOR THE WAVE-PARTICLE DUALITY OR BUT FOR ALMOST ALL THE QUANTUM PHENOMENA.

THE PHASE ANGLE INDICATES THE ANGULAR POSITION OF THE NUCLEUS OF MASS AND CHARGE IN THE PHOTON. THEREFORE, ALL THE QUANTUM PHENOMENA ARE THE NATURAL PROOFS OF THE PRESENCE OF A NUCLEUS OF MASS AND CHARGE IN A PHOTON ACCORDING TO THE NEW QUANTUM THEORY.

The changes in the values of Sin θ and Cos θ with the rotation of the nucleus of mass around the centre of the photon continuously vary the velocities and accelerations of the photon in both the X and Y – directions and move the photon as a wave with UP and DOWN movements.

When the value of Cos θ is positive, the photon moves
:   UP

When the value of Cos θ is negative, the photon moves
:   DOWN

When the value of Sin θ is positive, the photon moves :
    FASTER in X – Axis from average velocity of Light 'c' meter per second

When the value of Sin θ is negative, the photon moves:
    SLOWER in X – Axis from average velocity of Light 'c' meter per second

WAVE-PARTICLE DUALITY

The above explains WHY a photon always move as a Wave?

A wave formed by a photon is shown in the **Figure 8**. The Figure shows the position of a photon at different phase angles of the nucleus of the photon in one wave cycle along with the directions of the applicable forces in X and Y directions.

The size of the arrows of the forces roughly indicates the magnitudes of the forces. The **Figure 8** is only indicative and 'Not To Scale'.

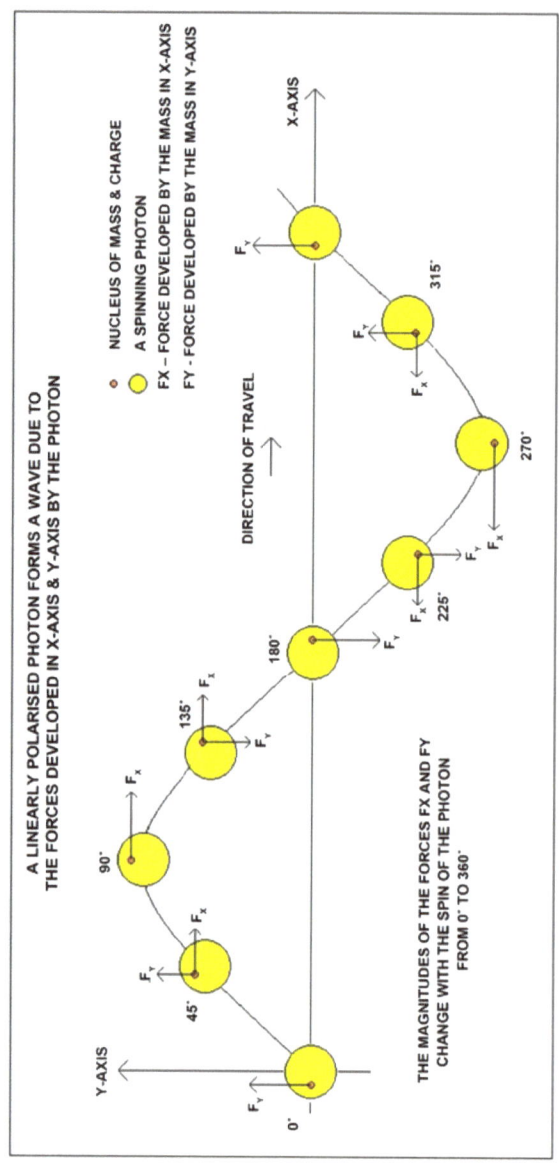

Figure 8: A photon forms a wave due to the 'Forces' developed by the mass in the nucleus of the spinning photon. The Figure shows the Forces at different phase angles.

NOT TO SCALE

HOW A PHOTON FORMS A WAVE

The following description explains how a photon forms a wave?

FORMATION OF 1ST QUARTER OF WAVE: (PHASE ANGLE FROM 0^0 TO 90^0)

The starting point of a new-wave cycle is 0^0 phase angle and the photon is at X – axis. In this quarter the photon spins and moves as a wave from phase angle 0^0 to 90^0 of the nucleus of the photon (from the point A to point B in **Figure 8**).

- The value of Sin θ increases continuously from the zero value at 0^0 phase angle to +1 at 90^0 phase angle. Therefore, the velocity and acceleration of the photon in the horizontal direction increase continuously in the 1st Quarter of Wave.

- At $0°$ phase angle, the origin of new-wave cycle, the value of Sin θ is zero, therefore the horizontal velocity of the photon (due to the mass in the nucleus of the spinning photon) is zero. At $0°$ phase angle, the photon moves in horizontal direction with 'c', the velocity of light.
- At $90°$ phase angle, the value of Sin θ is +1, which is maximum value. Therefore, the velocity of the photon in the horizontal direction (due to the mass in the nucleus of the spinning photon) is the maximum and the photon moves with the fastest velocity in the horizontal direction.

Horizontal velocity in X - direction at $90°$ phase angle: $c + 2\pi r f$ meter/sec

- The value of Cos θ decreases continuously from +1 at $0°$ phase angle to zero at $90°$ phase angle. Therefore, the velocity and acceleration of the photon in the vertical direction decrease continuously in the 1st Quarter of Wave.

- At 0° phase angle or the origin/starting point of a new-wave cycle, the value of Cos θ is +1 which is the maximum value of Cos θ. Therefore, the UPWARD vertical velocity of the photon is the maximum at the point A.

Vertical velocity in Y - direction at 0° phase angle: $2\pi r f$ meter/sec

- At 90° phase angle, the value of Cos θ is zero, both the velocity and acceleration of the photon in the vertical direction are the zero.
- In the 1st Quarter of Wave, the value of Cos θ is positive, therefore the photon moves up from the origin above the centreline of the photon travel.
- At 90° phase angle with Cos θ being zero, the upward velocity in Y - direction of the photon attains zero value, therefore the photon cannot go up further.
- At 90° phase angle, the photon reaches the highest point 'Crest' of the wave.

FORMATION OF 2ND QUARTER OF WAVE:
(PHASE ANGLE FROM 90⁰ TO 180⁰)

In this quarter the photon spins and moves as a wave from phase angle 90^0 to 180^0 of the nucleus of the photon (from the point B to point C in **Figure 8**).

- The value of Sin θ decreases continuously from the +1 at 90^0 phase angle to zero value at 180^0 phase angle. Therefore, the velocity and acceleration of the photon in the horizontal direction decrease continuously in the 2nd Quarter of Wave.
- At 90^0 phase angle, the 'Crest' of the wave cycle, the value of Sin θ is +1 with the maximum horizontal velocity of the photon (due to the mass in the nucleus of the spinning photon).
- As the photon spins and move from 90^0 phase angle, the value of Sin θ decreases continuously.

Therefore, the photon decelerates continuously from phase angle 90^0 to 180^0 in the horizontal direction.

- At 180^0 phase angle, the value of Sin θ is zero. Therefore the velocity of the photon in the horizontal direction (due to the mass in the nucleus of the spinning photon) is also zero and the photon moves in horizontal direction with 'c', the velocity of light.

- From the phase angle 90^0 to 180^0, the value of Cos θ decreases continuously from the zero value at 90^0 phase angle to $(-)1$ at 180^0 phase angle.

- The value of Cos θ being negative in the 2nd Quarter, the photon moves DOWNWARD during the phase angle 90^0 to 180^0 to form the wave.

- As the negative value of Cos θ increases continuously during the phase angle 90^0 to 180^0, the velocity and acceleration of the photon in the DOWNWARD vertical direction increase continuously in the 2nd Quarter of the Wave.

- At 180° phase angle, the value of Cos θ is (-)1 which is the minimum value of Cos θ (or the maximum value in the reverse direction: DOWNWARD). Therefore, the DOWNWARD vertical velocity of the photon is the maximum at 180° phase angle.

Vertical velocity in Y - direction at 180° phase angle: (-) 2 π r f meter/sec (DOWNWARD)

- In the 2nd Quarter of Wave, the value of Cos θ is negative, therefore the photon moves DOWN from the 'Crest' point above the centreline of the photon travel and at 180⁰ phase angle, the photon reaches the centreline of the wave.

WAVE-PARTICLE DUALITY

NOTE:

During the first half wave formation from 0^0 to 180^0 phase angle, the value of Sin θ is positive (zero or more), therefore, the velocity of the photon (due to the mass in the nucleus of the spinning photon) in the horizontal direction ($2\pi rf \sin \theta$) is positive. This adds to the normal velocity of photon 'c'.

During the first half wave formation, the photon moves faster in its direction of travel, the horizontal direction ($c + 2\pi rf \sin \theta$), than the normal velocity of light 'c'.

FORMATION OF 3RD QUARTER OF WAVE:
(PHASE ANGLE FROM 180^0 TO 270^0)

In this quarter as the photon spins and moves as a wave from phase angle 180^0 to 270^0 (from the point C to point D in Figure 8).

- The value of Sin θ decreases continuously from the zero value at 180^0 phase angle to (-)1 at 270^0 phase angle. Therefore, the velocity and acceleration of the photon in the horizontal direction decrease continuously in the 3rd Quarter of the Wave.
- At 180^0 phase angle, the photon is at the centreline of the wave, the value of Sin θ is zero, therefore the horizontal velocity of the photon (due to the mass in the nucleus of the spinning photon) is zero and at 180^0 phase angle, the photon moves in horizontal direction with 'c', the velocity of light.

- At 270^0 phase angle, the value of Sin θ is (-)1, which is minimum value (or the maximum value in the REVERSE direction).
- The velocity of the photon in the horizontal direction (due to the mass in the nucleus of the spinning photon) is the maximum in the REVERSE direction and the photon moves with the slowest velocity in the horizontal direction.

Horizontal velocity in X - direction at 270^0 phase angle: $c - 2\pi r f$ **meter/sec**

- The value of Cos θ increases continuously from (-)1 at 180^0 phase angle to zero value at 270^0 phase angle. Therefore, the velocity and acceleration of the photon in the vertical direction increase continuously (decreases continuously in the DOWNWARD direction) in the 3rd Quarter of Wave.

- At 270° phase angle, the value of Cos θ is zero, both the velocity and acceleration of the photon in the vertical direction attain the zero value.
- In the 3rd Quarter of Wave, the value of Cos θ is negative, therefore the photon moves DOWN from the centreline at 180° phase angle to below the centreline of the photon travel.
- At 270° phase angle the value of Cos θ is zero, therefore the DOWNWARD velocity in Y - direction of the photon attains zero value and the photon cannot go down further.
- At 270° phase angle, the photon reaches the lowest point 'Trough' of the wave.

Vertical velocity in Y - direction at point D: 0

WAVE-PARTICLE DUALITY

FORMATION OF 4TH QUARTER OF WAVE: (PHASE ANGLE FROM 270^0 TO 360^0)

In this quarter as the photon spins and moves as a wave from phase angle 270^0 to 360^0 (from the point D to point E in Figure 8.)

- The value of Sin θ increases continuously from the (-)1 at 270^0 phase angle to zero value at 360^0 phase angle. Therefore, the velocity and acceleration of the photon in the horizontal direction increase continuously in the 4th Quarter of the Wave.
- At 270^0 phase angle, the 'Trough' of the wave cycle, the value of Sin θ is (-)1 with the minimum horizontal velocity of the photon (due to the mass in the nucleus of the spinning photon).
- As the photon spins and move from 270^0 phase angle, the value of Sin θ increases continuously.

Therefore, the photon accelerates continuously from phase angle 270^0 to 360^0 in the horizontal direction.

- At 360^0 phase angle, the value of Sin θ is zero. Therefore the velocity of the photon in the horizontal direction (due to the mass in the nucleus of the spinning photon) is also zero and the photon moves in horizontal direction with 'c', the velocity of light.

- From the phase angle 270^0 to 360^0, the value of Cos θ increases continuously from the zero value at 270^0 phase angle to +1 at 360^0 phase angle.
- The value of Cos θ being positive in the 4th Quarter, the photon moves UPWARD during the phase angle 270^0 to 360^0 to form the wave.
- As the positive value of Cos θ increases continuously during the phase angle 270^0 to 360^0, the velocity and acceleration of the photon in the UPWARD vertical direction increase continuously in the 4th Quarter of the Wave.

- At 360° phase angle, the value of Cos θ is +1 which is the maximum value of Cos θ. Therefore, the UPWARD vertical velocity of the photon is the maximum at 360° phase angle.

Vertical velocity in Y - direction at 360° phase angle: $2\pi r f$ meter/sec (UPWARD)

- In the 4th Quarter of the Wave, the value of Cos θ is positive, therefore the photon moves UP from the 'Trough' point below the centreline of the photon at 270° phase angle and the photon reaches the centreline of the wave at 360°.
- This completes one full wave cycle from 0° to 360° and a new wave cycle starts.

NOTE:

During the second half wave formation from 180^0 to 360^0 phase angle, the value of Sin θ is negative (zero or less), therefore, the velocity of the photon (due to the mass in the nucleus of the spinning photon) in the horizontal direction is negative. This reduces the velocity of the photon from its normal velocity 'c'.

During the second half wave, the photon moves slower than the velocity of light 'c'.

In the full wave cycle of the photon from 0^0 to 360^0, the increase in the horizontal velocity during the first half wave cycle is nullified by reduction in velocity during the second half wave cycle. Therefore, the photon travels with average velocity 'c', the velocity of light.

WAVE-PARTICLE DUALITY

A PHOTON FORMS A WAVE DUE TO THE FORCES OF VARYING INTENSITY AND DIRECTION DEVELOPED BY THE MASS IN THE NUCLEUS, IN OFF-CENTRE POSITION, OF THE PHOTON.

If a photon does not have a nucleus of mass in off-centre position, it can never display the Wave-Particle Duality phenomenon.

Wave-Particle Duality phenomenon proves the existence of the mass in off-centre position of a photon and is also a natural proof of New Quantum Theory.

CHAPTER 4.3

TO FIND EXACT LOCATION OF A PHOTON IN A WAVE CYCLE

The wave profile of a photon is not sinusoidal. A photon has its own wave profile due to the continuously varying acceleration or deceleration due to the mass in the nucleus of the spinning photon.

WAVE-PARTICLE DUALITY

The velocities and accelerations of a photon at any position in a wave cycle between $0°$ to $360°$ (θ phase angle of the nucleus/photon) are already calculated in Chapter 4.1 and given below:

- Horizontal velocity in X-direction: $c + 2\pi r f \sin\theta$
- Vertical velocity in Y-direction: $2\pi r f \cos\theta$
- Acceleration a_x in X - Direction:

$$[2\pi r f \sin(\theta + d\theta) - 2\pi r f \sin\theta] / dt \text{ m/sec}^2$$

- Acceleration a_y in Y - Direction:

$$[2\pi r f \cos(\theta + d\theta) - 2\pi r f \cos\theta] / dt \text{ m/sec}^2$$

A photon of frequency 'f' spins or rotate $360°$ 'f' number of times in one second.

$$\text{Time to rotate } 1° \text{ by a photon} = \frac{1}{360 f} \text{ second}$$

TO FIND THE EXACT LOCATION OF THE PHOTON IN A WAVE CYCLE AT ANY PHASE ANGLE OF THE PHOTON USE THE FOLLOWING EQUATIONS:

$$S = ut + 1/2at^2$$

As the photon spins a fractional angle $d\theta$ from phase angle θ to phase angle $(\theta + d\theta)$ in time dt, the photon travels the fractional horizontal distance dS_H and the fractional vertical distance dS_V.

HORIZONTAL DISTANCE travelled by the photon during the spin of the photon from (θ) to ($\theta + d\theta$) in time dt:

$$dS_H = [c + 2\pi rf \sin\theta]\, dt + [\pi rfm \{\sin(\theta + d\theta) - \sin\theta\}]\, dt^2$$

The Integration of the above equation with the limits of θ phase angles calculates the horizontal distance covered by the photon in a wave cycle.

The horizontal distance travelled by a photon during 1st Quarter of the wave from $0°$ to $90°$ is given by the following equation:

$$S_H = \int_0^{90} [c + 2\pi r f \sin\theta]\, dt + [\pi r f m \{\sin(\theta + d\theta) - \sin\theta\}]\, dt^2$$

The horizontal distance travelled by a photon during 1st Half of the wave from $0°$ to $180°$ is given by the following equation:

$$S_H = \int_0^{180} [c + 2\pi r f \sin\theta]\, dt + [\pi r f m \{\sin(\theta + d\theta) - \sin\theta\}]\, dt^2$$

The above is the horizontal distance covered by the photon during its first half spin of the photon from $0°$ to $180°$.

DURING THE FIRST HALF SPIN OF THE PHOTON WITH ITS NUCLEUS ROTATING FROM $0°$ TO $180°$, THE MASS IN THE NUCLEUS INCREASES THE HORIZONTAL VELOCITY OF THE PHOTON.

THEREFORE, THE HORIZONTAL DISTANCE COVERED BY THE PHOTON (FIRST HALF WAVELENGTH OF THE PHOTON) IS MORE THAN THE HALF OF THE WAVELENGTH OF THE PHOTON.

The horizontal distance travelled by a photon during 3rd Quarter of the wave from 180^0 to 270^0 is given by the following equation:

$$S_H = \int_{180}^{270} [c + 2\pi r f \, \text{Sin} \, \theta] \, dt + [\pi r f m \{\text{Sin}(\theta + d\theta) - \text{Sin} \, \theta\}] \, dt^2$$

The horizontal distance travelled by a photon during 2nd Half of the wave from 180^0 to 360^0 is given by the following equation:

$$S_H = \int_{180}^{360} [c + 2\pi r f \, \text{Sin} \, \theta] \, dt + [\pi r f m \{\text{Sin}(\theta + d\theta) - \text{Sin} \, \theta\}] \, dt^2$$

The above is the horizontal distance covered by the photon during its second half spin of the photon from 180^0 to 360^0.

WAVE-PARTICLE DUALITY

DURING THE SECOND HALF SPIN OF THE PHOTON WITH ITS NUCLEUS ROTATING FROM 180^0 TO 360^0, THE MASS IN THE NUCLEUS DECREASES THE HORIZONTAL VELOCITY OF THE PHOTON.

THEREFORE, THE HORIZONTAL DISTANCE COVERED BY THE PHOTON (SECOND HALF WAVELENGTH OF THE PHOTON) IS SHORTER THAN THE HALF OF THE WAVELENGTH OF THE PHOTON.

VERTICAL DISTANCE travelled by the photon during the spin of the photon from (θ) to ($\theta + d\theta$) in time dt:

$dS_V = 2\pi r f \cos\theta\, dt + \pi r f m\, [\cos(\theta + d\theta) - \cos\theta]\, dt^2$

The Integration of the above equation with limits of phase angles provides the Vertical Distance travelled by a photon in a wave cycle.

The vertical distance travelled by a photon during 1st Quarter of the wave from 0^0 to 90^0 is given by the following equation:

$$S_V = \int_0^{90} 2\pi r f \cos\theta \, dt + [\pi r f m \{\cos(\theta + d\theta) - \cos\theta\}] \, dt^2$$

THE ABOVE VERTICAL DISTANCE DURING 0 TO 90^0 PHASE ANGLE IS THE AMPLITUDE OF THE WAVE.

The vertical distance travelled by a photon during 3rd Quarter of the wave from 180^0 to 270^0 is given by the following equation:

$$S_V = \int_{180}^{270} 2\pi r f \cos\theta \, dt + [\pi r f m \{\cos(\theta + d\theta) - \cos\theta\}] \, dt^2$$

THE ABOVE DISTANCE IS THE AMPLITUDE OF THE WAVE IN THE SECOND HALF CYCLE OF THE WAVE. THE VALUES OF BOTH THE AMPLITUDES IN THE FIRST AND SECOND HALF ARE EQUAL.

www.ingramcontent.com/pod-product-compliance
Lightning Source LLC
Chambersburg PA
CBHW040321220526
45473CB00009B/2515